My Adventures in Woodcutting

By Brian Tate

A selection from *The World Is a Handkerchief: Thirty Short Stories by People Like You!*

Edited by Anthony W. Parr

The World Is a Handkerchief: Thirty Short Stories

Copyright © 2022 by Anthony W. Parr, Ginkgo Leaf Publishing

All rights reserved. No part of this book may be reproduced in any form or by any electronic or mechanical means, including information storage and retrieval systems, without permission in writing from Ginkgo Leaf Publishing, except by a reviewer, who may quote brief passages in a review.

All illustrations and photographs are used with permission of the artists and photographers, who retain their copyrights.

All illustrations copyright © Anthony W. Parr, unless otherwise indicated.

Cover and book design by Anthony W. Parr.

Ginkgo Leaf Publishing

PO Box 6217, Bellevue, WA 98008

425-444-1656

Dedication

To all the authors who shared their eclectic stories,

This is for you.

To all the readers with great expectations,

This is for you.

Enjoy.

Art
"Art washes away from the soul
the dust of everyday life."
Pablo Picasso

Travel
"Like all great travelers, I have seen more than I remember, and remember more than I have seen."
Benjamin Disraeli

Nature
"I like this place and could willingly waste my time in it."
William Shakespeare

And...life!
"You only live once, but if you do it right, once is enough."
Mae West

PRAISE FOR

THE WORLD IS A HANDKERCHIEF

and

MY ADVENTURES IN WOODCUTTING

I am grateful for your wonderful book - just reading about the woodcutter from Lake Wilderness. Emerson once wrote: "Live in the Sunshine, Swim in the Sea, Drink the wild Air." When you share stories with us, I feel as though we all get to partake in Emerson's world!

--Lori Carmody

MY ADVENTURES IN WOODCUTTING

By Brian Tate

A Woodpecker enjoys the woods as much as Brian does.
~Anthony W. Parr

VERY EARLY IN LIFE I developed a keen interest in self-reliance and sharing that sort of experience and knowledge with others. I didn't realize it at the time, but my father helped stoke the fires of my passion for woodcutting through his side job selling chainsaws.

My dad sold Pioneer chain saws. This was a highly respected brand. As a teenager, chain saws were something I could really get into. I do not remember the exact year, but one Christmas I received a shiny new Pioneer chain saw which changed my life forever.

I also developed an interest in making money so I could acquire more shiny new stuff. That was not an obsession with me and never has been, but still, it is a lot of fun to work hard and acquire enjoyable things from time-to-time.

Once I got my driver's license, I was able to expand my radius of travel to find places to cut firewood for our family. I quickly discovered that I could cut wood and sell it.

OUR FAMILY OWNED a heavily-wooded lot on Lake Wilderness in Maple Valley, about 15 miles from our home. It seemed that every year a couple of trees came down in wind storms. Also, during those years there was a great deal of building being done on the east side of Lake Sammamish. Builders were usually happy to push logs out to the side of the lots and have young entrepreneurs like my friends and me pick up the wood.

We had to travel a little distance, but the Weyerhaeuser lumber mill in Snoqualmie was a delightful place to visit. They would cull quite a few logs and put them in a big pile on the corner of their property. Many were fairly large logs. My friends and I had chain saws that were a little bigger than most homeowner size saws, which cut out quite a bit of competition for those Weyerhaeuser logs; they were simply too big for most people to make use of.

There was somewhat of a cult following of Pioneer chain saws, but they were not worth a great deal. One day I made the decision to sell everything Pioneer that I had. I hoped to raise enough money to buy two new Stihl chain saws. I wanted a good size saw with about a 28-inch bar for cutting up relatively large logs and a smaller saw with a 16-inch bar for bucking up firewood.

Eventually I acquired a seven-acre property in Snoqualmie, badly overgrown with many big logs already down. A wood cutter's paradise! The house had a large fireplace with a large insert that would easily heat the whole house, even though it was built in 1950 with poor insulation. Burning firewood saved us a lot in heating expenses.

It has always felt to me like wood heat warms clear to the bone while other sources of heat just sort of warm a body on the surface.

WHEN I PURCHASED the Snoqualmie house, there was an old Farmall Model A tractor on the property. The seller didn't want it so I paid a little extra cash for it. That tractor proved to be very useful in pulling logs out of little swales and other tricky spots where it would be difficult to cut the logs. Using that tractor, another passion was sparked.

The Weyerhaeuser cull pile remained available for a couple years more. During those years my little farm's trees and Weyerhaeuser's cull pile provided plenty of wood, but there wasn't any left over to sell.

At this time, a long-held dream to become a volunteer firefighter became a reality. I approached my Snoqualmie neighbor, had a few discussions with him about being a volunteer, and instantly signed on. In that capacity, we made a few visits to the Weyerhaeuser lumber mill for inspections. It was fascinating. Those visits planted another seed: I imagined that there could be an alternative and greater use for some of my logs in the future.

As the years passed and I got close to retirement age, I realized that there was cross-pollination between many

of the things – including wood cutting and tractors – I had passions for.

FOR EXAMPLE, I LEARNED that tractors historically played a vital role in moving and processing wood, and still do today. Many farmers had what is commonly called a "buzz saw" on the back of a tractor, driven by a leather belt, or PTO (power take off). A buzz saw can be just about any circular saw; the name comes from the noise they make. Saws on the back of a tractor were also called cord wood saws, saw rigs, and sawing machines.

I joined a small club of tractor enthusiasts and discovered that a couple of guys in our group had their own sawmills. One of them had a really neat buzz saw on the back of one of his 1940s tractors that he would take to shows. Our club toured a couple of these small sawmills and I starting thinking about taking some of my own logs to one of them.

I discovered there are actually quite a few large saws and planers east of Lake Washington. Mostly people have acquired them at little or no cost from mills that went out of business. They got them for their own use and perhaps for a few friends and family. However, projects like these are not for the faint of heart. They take up a fair amount of space and need to be covered against weather. On average, a fairly large saw or planer needs a building or at least a pole barn roof about 30 feet by 80 feet. Plus, these machines require a fair amount to power. You don't just plug them into a 110-volt outlet. A large tractor and other equipment are required to handle the logs.

Getting Logs to a Sawmill

Duane Isaksson lives in Redmond, Washington. He still operates a sawmill that his father started about 75 years ago. Their equipment has come from other mills in the area that have shut down. Duane operates his 48-inch circular saw and eight-foot planer with large diesel engines. These engines require a fair amount of maintenance, especially because of their age. They were built in the USA by General Motors and sent to Germany to power tanks in WWII. Duane modified the engines to use only four of their eight cylinders.

I asked Duane why he does not use electric motors. The mill is located just off a major state highway, so electric lines are nearby. I do not know a lot about electricity, but these machines would probably need electric motors requiring 440 volts. Duane told me the installation would be expensive even though the lines are close by. Additionally, the power company has a minimum monthly charge for that kind of power usage. Duane would have to pay for a minimum amount of power even if he did not use it.

I think old diesel engines are cool in their own special way. And Duane seems to like the idea of self-reliance.

I have had many adventures harvesting logs and getting them to the mill. Every year we have a couple trees come down at our family property at Lake Wilderness.

Our lot is fairly steep, which presents some challenges when harvesting logs. The top of the property is not too bad. I have had to pull a few logs up a trail with a big switchback. To do this, I attached a snatch block to a big tree just beyond the corner of the switchback. (A snatch block is a heavy-duty pulley inside a metal casing that lets you to change the direction of a winch's cable by off-setting the anchor point, allowing you to pull in a three-point motion. The snatch block I use came from Young Iron Works in Seattle, started in 1902 and still in business today.) I attached a cable to a log, threaded the cable through the snatch block and finally attached the other end of the cable to my pickup truck. Once everything was in place I simply drove forward with my pickup and pulled the log up to the snatch block. Once the log reached this point, I shortened the cable and reattached it to my pickup truck. I then drove again, pulling the log the rest of the way up the hill to our landing (parking) space.

THE NEXT STEP is getting the log onto a vehicle for transport to the sawmill. In big logging operations they used a spar pole. This was often a live tree that was limbed, or sometimes an already-harvested tree that was stood up vertical and supported with at least three cables anchored to the ground.

Spar poles were useful in pulling logs out of the forest, much like the operation I described using my pickup. They were also used to lift logs and put them on the bed of a truck or trailer. In later years these spar poles evolved into big pieces of steel equipment called yarders, with the tree replaced by a sturdy steel pole.

Getting a good-sized pulley or snatch block up to the top of a tree in order to lift logs onto a truck can be a time consuming, difficult and dangerous job. This was not practical for my work on the family property. Instead, I have an 18-foot trailer that I use for transporting logs to the mill. My trailer has a tilt bed and I recently mounted a winch on the front. This makes it pretty easy to pull the logs onto the trailer. There have been other times when I have used my pickup truck to get logs to the mill. If I have a bunch of logs, I might use both my pickup truck and the trailer.

I have a fairly new Kubota four-wheel-drive tractor with a front loader that I use for loading logs onto my pickup truck. This isn't always as easy as it may sound. Logs are heavy and often I cannot lift the whole log with my tractor. In this case I have to place one end of the log onto the back of the truck, then reposition the tractor to the other end of the log to lift and push it into the truck bed.

I MENTIONED THAT OUR property at Lake Wilderness is pretty steep. We have trees that fall near the bottom of the lot, near the lake. It would be challenging and time consuming to pull them up the hill to the parking area, so with those lower logs I have pulled them down the hill and into the lake. It is much easier to get a log down a hill than it is to get it up a hill! Once in the water, I lashed several logs together into a small raft. I pulled them a short distance to a boat launch and floated them into my trailer. This process takes time and effort, but is kind of fun.

I recently learned that some 75 years ago, there was a sawmill at our property at Lake Wilderness. The city park has some pictures of it. For the 50-plus years my folks have owned their lot, there have been some logs submerged in the lake. A couple of years ago, with the help of my brother and some friends, we pulled out some of those logs to make the lake more inviting and safer. I cut them up for firewood. This past summer, I pulled a bigger log out of the lake, a Douglas fir, roughly 28-30 inches in diameter and about ten feet long. It was waterlogged and very heavy. I broke a cheap come-along trying to get it up on the beach. I have cut a couple thin slices off the ends and it smells wonderful, like the day it was originally cut down. Wood generally stays in good condition when it is completely submerged under water. I think this log will make some nearly perfect vertical grain lumber. It is so heavy, though, I am going to have to let it dry out for at least a year before I can move it. I am trying to decide if I want to winch it to the top of the hill, about 200 feet and pretty steep, or tow it across the lake to the boat launch.

Getting logs onto a truck or trailer is a really good feeling. That marks the end of the really hard work.

WHEN I ARRIVE his mill, my friend Duane has a good-sized tractor for unloading the logs. He uses a set of tongs that look just like a giant set of salad tongs attached to the front loader of his tractor. The ends of the tongs are very sharp. The tongs are designed to open up to 20 inches for larger logs and stumps. Once closed onto a log, the harder you pull, the tighter the tongs sink into the bark and wood for dragging and skidding. Tongs are very useful in all kinds of log handling, both in the woods and at the mill.

Duane has a large flat deck that feeds directly into his saw, powered by an electric motor and chain. Sometimes he puts incoming logs directly on the deck, or, during busier times when the deck is filled up, he makes rows of logs off to the side of the sawmill.

Visiting Duane at the Isaksson Mill is a real treat, whether a person is actively doing business there or just touring and learning about this historic sawmill. Duane has hosted many groups from museums, granges, and various clubs. There are a number of online links to the Isaksson Mill.

Duane and Brian at the sawmill.

Discovering a Log's Best, Highest Use

IN THE LAST FEW YEARS, I have had more time to cut wood. I love the exercise and it takes some engineering skills to take trees down in a safe and efficient manner. Also, some engineering analysis can make splitting a particular piece of wood much easier. These days, though, it can be a burden to get rid of downed or unwanted trees. Fewer people want the wood as fewer homes use wood for heat. For these reasons friends often call me and ask if I want the wood. I have plenty of wood for myself and selling firewood doesn't really appeal to me anymore. I wondered what else I could do with the wood and remembered my friends with sawmills. I like to see materials go to their highest and best use, so when I have good looking logs available, I take them to the sawmill where they can make boards from the best portion and I cut the rest up for firewood. It is an approach that works very well for me.

Thinning Douglas Fir trees.

This past summer, a friend called and asked if I wanted the logs from trees that she had recently taken down. I said I would like them. There were quite a few logs, so this took several days.

A neighbor and my friend Tony became interested in the project. Tony ended up going to the mill with me and we had some big live edge planks made for him that could be used as benches in his garden.

Not all logs get to go to the mill. There are a number of criteria. Duane likes them to be at least eight feet long and I believe the maximum length is about 24 feet. They also have to be big enough in diameter, probably at least eight inches. They can have much larger diameter, but obviously there is a limit. In general, any trees currently growing around western Washington are OK. Most of the logs that come in are 30 to 40 inches in diameter.

There are many species of wood in this area. I recently asked Duane if I could bring in a couple of madrone logs. He said, "No! Cut them up for firewood." I thought they might make really interesting boards, cuts that Duane usually makes, such as dimensional lumber and live edge planks. He said he did some madrone once and it warped badly.

Often people let logs sit around, trying to decide what to do with them. By the time they decide to try to sell them, or more likely give them away, they are too rotten. A little bit of aging in the wood and sometimes even a little rot can make some really interesting product. A person would not want to build a house out of such wood, but it can be very interesting for smaller projects, like furniture or art work.

DUANE IS A VERY WARM person who loves to tell stories about the many things he has experienced in the sawmill business.

My favorite story involves one of his customers who works with a big commercial tug boat company.

Duane in his sawmill.

This person frequently encounters a variety of logs. Sometimes they are just individual logs that are floating around in the ocean and look interesting to him. They are often large logs, but he has all the necessary equipment to easily harvest and transport them. Some of these floating logs have had metal of various types put into them, or they have had rocks and dirt beat into them by various means.

Any type of foreign object can be very damaging to the saw blade or even dangerous, just like foreign objects being introduced into an aircraft propeller or engine. The saw blade is so big – 48 inches – that individual teeth can be replaced when needed, but that process is time consuming (down time), expensive and irritating to the saw operator.

Duane said that he told this customer, "There is a reason these logs are out floating around in the ocean." He went on to explain it is probably because no one else wanted them.

I have a pretty clear idea of the criteria for Duane's mill, so I'm glad to say he has never sent me home with any logs. There have been a couple of times, though, where there was rot in the middle of a log, making some of the finished product not very useful for anything other than firewood.

One of my many other hobbies in the past (also related to being self-reliant) has been raising my own beef and pork. Taking logs to Duane is just like giving cutting instructions to a butcher.

THERE ARE MANY THINGS that can be done with the logs. It can simply be rough cut into dimensional lumber (cut on all four sides), or a person can have what they call live edge or wild edge on one or more sides of a board (left in their natural state with bark still on the edges). Logs can be cut to just about any size. I once had Duane cut me a big beam that was 12 inches x 12 inches x eight feet (Duane's minimum length).

Like just about any other discipline, there is some unique vocabulary. Much of the rough-cut wood I have gotten is "five-quarter." That means it is five-quarters-of-an-inch thick. In other disciplines, people would say something is an inch-and-a-quarter thick. I have no idea how this unique terminology came about. And there are many other terms. Duane often gets requests for vertical grain. This means the growth rings of the log or board are vertical to the face of the board. And then there are synonyms: vertical grain is really the same as rift cut or quarter sawn.

Many customers just get the rough-sawn wood from Duane. In this wood, the circular saw marks are clearly visible. Duane tells me this is a trending thing right now, to see those circular saw marks. Most commercially-cut wood is cut with big band saws and you get a much different look; it doesn't have much character.

Duane has a large planer, driven by one of his big diesel engines. A planer is a machine that makes the wood smooth on one or more sides. There are also special knives and settings that can be used to make some interesting products, like shiplap. Shiplap was once a very popular shape of wood. It was used for many years as siding on houses and seems to be making a big comeback. There is a restaurant near my house and all of the seating benches are made out of reclaimed shiplap. Every time I go in there, I tell the young workers what shiplap is. They generally have no idea and really are not interested.

I recently had someone give me some big cedar logs. This was the first time I took cedar to Duane, and the first time I had anything planed. It turned out really beautiful, and I am planning to use it for a small deck this summer.

I HAVE HAD MANY FUN experiences with woodcutting. I recently told a long-time friend that I had some wood milled by Duane. He told me that years ago he also had some wood milled by Duane. When I was in high school, I helped this friend clear the land for his house, which is near Duane's sawmill. The logs we cleared probably went to Duane, but I had no idea at the time.

Just the other day Duane and I were talking, and he mentioned a mutual friend who got most of the wood for his house milled by Duane several decades ago.

Brian using a tractor to load logs into his pickup.

I recently bought a cabin in South Cle Elum, Washington, about 70 miles east of my house. I made an interesting friend there this past summer. He owns several historic buildings in town, which was incorporated in 1902. One of them is the former South Cle Elum City Hall. He was showing me this building. He had a large stack of lumber that Duane had milled for him. He is going to use this lumber for further restoration of the building.

My daughter recently bought a house in Ellensburg, Washington, about 90 miles from my home in Snoqualmie. We didn't realize it at the time, but there was a big birch tree that had been dead for some time on her lot. A good friend of mine who enjoys climbing trees and has all the necessary gear to climb and safely limb and cut trees down in pieces helped us get it down. We haven't really made anything with the wood, but we did get an interesting mix of dimensional lumber and live edge, including one slab that is four inches thick. This was the first time I had taken a birch tree to Duane's mill. Another eight-foot-long birch log became available in Ellensburg, and it is sitting at the mill waiting to be cut.

When I was young, I had relatives I visited who lived in Moses Lake, some 150 miles to the east. I made friends with a few people there. As luck would have it, one of those friends recently cut down some trees that were planted by his parents at least 50 years ago. They were special trees, apricot and maple. As I told my friend about my recent adventures with the sawmill, he became excited about the possibility of getting a few of his logs to the mill. It was a long way to haul these logs, and my friend is not able to drive much, so I took the logs to Duane's mill. My friend ended up with mostly live edge from the logs.

Duane using tongs to move logs.

When you make square lumber out of round logs you end up with a fair amount of trimmings. Sometimes interesting things can be done with trimmings. After my friend's logs were cut, I loaded up the milled wood and most of the trimmings and took them back to Moses Lake. I am anxious so see what they do with their wood.

Several years ago, I met a guy in Ellensburg who goes to craft events selling wonderful hand-carved wood toy trucks, tractors, and other items. He is an older veteran and he donates several of his pieces each year to charity fundraisers. I love what he does and have donated lots of wood to him. Every now and then he gifts me a finished toy as a thank you, but mostly I buy his carvings like everyone else. It's hard to describe why I like them so much. I do enjoy sitting in a nice comfortable room, looking at the wood that I harvested or reclaimed, seeing what creative thing someone made from it while remembering that sometimes I was up to my waist in mud and freezing to death gathering that wood.

WOODCUTTING HAS BEEN a very educational and rewarding adventure for me. I love to be outside and enjoy the physical exercise that is required to cut and gather wood. I have been cutting firewood for a very long time, but taking wood to the mill is a more recent and enjoyable endeavor.

Most of the wood I have had milled has dried very nicely and is ready to be worked. I have a small planer that I have used to smooth a few boards, making a couple of small book cases and things like that.

Afterword *by Anthony W. Parr:* **It's easy to see that Brian loves wood. I too, used to love sawdust up my nose. But Brian, loves helping people, assisting trees extend their life in becoming loving products. And he is a pleasure to know!**

Praise for

The World Is a Handkerchief

Of the 30 stories, I am especially enchanted by two about being a gardener. Growing up in Michigan, Colleen M. Donahue was surrounded by plants and gardens as a child. As an adult, managing the Master Gardeners' display and teaching garden in Bellevue, she has learned there is so much more to being a gardener. "Growing A Gardener" shares not only techniques, but also the many ways cultivating plants enriches the lives of all beings, especially humans. Maybin tells a similar story, but from a very different cultural starting point in Zambia. "Maybin's Garden" uses a conversational presentation style and includes vibrant drawings of the garden space. I found that both writers captured the essence and spirit of gardening in ways that were very relatable and inspiring.

--Brian Thompson, Manager of the Elisabeth C. Miller Horticultural Library, University of Washington Botanic Gardens

***

I loved reading about "the world is a handkerchief" explanation! I like this idea of threads interlining just as our lives do with other people.

I also loved reading about the flower displays! Maybe it'll inspire me to start my own flower garden! It was very nice reading how passionate the person telling the story was about it.

--Maria Jose Felix, Composer/Sound Designer at SkyRoseMusic

***

I really enjoyed reading Wenas in My Sketchbook. I was transported to the campground through the author's writing. The sketches accompanying the stories were also very enjoyable.

--Raghav Mehta, Bird Photographer and Audubon Trip Leader

What an interesting, vivid account [Three Lessons on the Path to Enlightenment] told with the ability of understanding and embracing the characters in their environments and landscape surroundings. Only an acute observer can describe the moment with clarity and wisdom.

--Elfi Rahr

PRAISE FOR THE WORLD IS A HANDKERCHIEF

***

My! Oh! My! What a gift!

I held the book close to my heart, a little emotional and then my heart got bubbly excited. Tony, your heart gift landed in two grateful hands and as expected, my heart smiled heartily. The journeys, the details, the experiences around the world are all blending together in my heart.

Why aren't they more human beings like Tony, I asked myself? Then the world will be far richer and better. No sooner than I asked that question, I heard a whisper in my heart. "The stories are saying look closely, there are many Tony's dressed in different garments. The book, the stories, are seeds Tony gathered along the way and freely shared to give evidence of deep, deep hope. One man's journey becomes the tillage of the heart from which all arts follow. It's like many incredible movies weaved together to make a masterpiece. Thank you, Tony, for gathering us together to shine a spotlight on the essence of humanity. You've handed a torch light to me, to go further in my life work of tending the garden. I shall hold soil and seed tightly close to my heart.

You've also given me new members of the family by weaving all these stories together. This summer I'll sit in the garden and read these stories to children and their parents. The thirty stories will then become many stories.

--Maybin Chisebuka, The Amazing Gardener

***

Tony, reading your Elvis story sparked my own fond memories of my mother and her love for Elvis, and my own love for her and his music. Through the years my flavor of rock and roll strayed a long way from what my mother would accept as music but we always could agree on Elvis. I got my guitar when I was nine and the first thing I did was run around the house imitating the king. My mom passed away of cancer in 2018 in what turned out to be a very sad chapter in my life. Shortly after she got sick, but had not told me yet, she sent me her original pressing of Blue Hawaii, all 14 songs. I can see her dancing in the living room of the old house every time I play this album. It still brings tears to my eyes as I write this but I will forever be grateful to my mom and Elvis for filling my heart with the love of rock and roll.

--Don Sanders, Investor, Musician and Desert Dweller

ABOUT THE EDITOR

Anthony W. Parr is an Author, Illustrator, and Artist.

This collection, The World is a Handkerchief, is the first book he has compiled and edited.

Somewhere near Casa Grande is his first Illustrated story book, with a series to come.

Many poems, eulogies, letters, and stories have flowed from Parr's pen.

His artwork includes pen drawings of many famous authors, in situ, at book events,

and charcoal drawings of theatrical artists and

acts in action.

Wide-ranging plein air drawings in the desert,

at the seaside,

and in the countrysides of England, Germany, France,

Italy, Australia, and the US.

Oil paintings and watercolors of grand visions in

Milan, Chaing Mia, and Athens.

OTHER BOOKS BY ANTHONY W. PARR

Somewhere Near Casa Grande

When a fox named Fresno encounters three border-crossing foxes from Mexico, they all wind up partying at a palm oasis—in a series of enchanting illustrations by British-born artist **Anthony W. Parr.** "A fox knows no borders, just how to have a good time in the desert," proclaims this parable, available now on Amazon in English and Spanish.

Parr's first book, *Somewhere Near Casa Grande,* shows his talent as a desert artist who has extensively hiked and painted near his part-time home in Palm Desert (he lives the rest of the year in Bellevue, Washington), as well as the backcountry near Tucson where his tale is set.

The book was inspired by his daughter **Kristina Parr** ("Why don't you write about foxes?") and is the first of several from his own Ginkgo Leaf Publishing company.

Parr tells readers of **California Desert Art** that he will send a free copy to teachers or anyone wishing to learn the fate of the furry comrades.

Email: tony@machinesandmethods.com

Review by Ann Japenga in California Desert Art Newsletter.

Made in United States
Orlando, FL
22 January 2023

28945981R00020